The Standard® AIRCRAFT LOG

ASA-SA-1

THE STANDARD® AIRCRAFT LOG
SA-1

ASA-SA-1
ISBN 978-1-56027-419-3

Published by
Aviation Supplies & Academics, Inc.
7005 132nd Place SE
Newcastle, Washington 98059 USA
Website: asa2fly.com
Email: asa@asa2fly.com

Printed in the United States of America

[19] 2025

TRANSPORTATION USD $9.95

Aircraft Record General Information

Manufacturer ______________________ **Model** ______________________

Serial ______________________ **Registration Number** ______________________

Date of Manufacture ______________________

Engine(s) currently installed:

Manufacturer ______________ **Model** ______________ **Serial** ______________

Manufacturer ______________ **Model** ______________ **Serial** ______________

Propeller(s) currently installed:

Manufacturer ______________________ **Model** ______________________

HUB Model ______________ **Serial** ______________ **Serial** ______________

Blade Model ______________ **Serial** __________ **Serial** __________ **Serial** __________

Blade Model ______________ **Serial** __________ **Serial** __________ **Serial** __________

____

____

____

Registered Owner Record

Name ______		Address ______	
City ______	State ______	From ______	To ______
Name ______		Address ______	
City ______	State ______	From ______	To ______
Name ______		Address ______	
City ______	State ______	From ______	To ______
Name ______		Address ______	
City ______	State ______	From ______	To ______
Name ______		Address ______	
City ______	State ______	From ______	To ______
Name ______		Address ______	
City ______	State ______	From ______	To ______

Notes

YEAR 20 _____ DATE	TOTAL TIME IN SERVICE	Current Weight and Balance Information			
		Empty Weight	Empty CG	Useful Load	Remarks

YEAR 20 ______ DATE	TOTAL TIME IN SERVICE	Current Weight and Balance Information			
		Empty Weight	Empty CG	Useful Load	Remarks

YEAR 20_____ DATE	RECORDING TACH TIME	TODAY'S FLIGHT	TOTAL TIME IN SERVICE	**Description of Inspections, Tests, Repairs and Alterations** Entries must be endorsed with Name, Rating and Certificate Number of Technician or Repair Facility. (See back pages for other specific entries.)

YEAR 20_____ DATE	RECORDING TACH TIME	TODAY'S FLIGHT	TOTAL TIME IN SERVICE	**Description of Inspections, Tests, Repairs and Alterations** Entries must be endorsed with Name, Rating and Certificate Number of Technician or Repair Facility. (See back pages for other specific entries.)

YEAR 20_____ DATE	RECORDING TACH TIME	TODAY'S FLIGHT	TOTAL TIME IN SERVICE	**Description of Inspections, Tests, Repairs and Alterations** Entries must be endorsed with Name, Rating and Certificate Number of Technician or Repair Facility. (See back pages for other specific entries.)

YEAR 20_____ DATE	RECORDING TACH TIME	TODAY'S FLIGHT	TOTAL TIME IN SERVICE	**Description of Inspections, Tests, Repairs and Alterations** Entries must be endorsed with Name, Rating and Certificate Number of Technician or Repair Facility. (See back pages for other specific entries.)

YEAR 20_____ DATE	RECORDING TACH TIME	TODAY'S FLIGHT	TOTAL TIME IN SERVICE	**Description of Inspections, Tests, Repairs and Alterations** Entries must be endorsed with Name, Rating and Certificate Number of Technician or Repair Facility. (See back pages for other specific entries.)

YEAR 20____ DATE	RECORDING TACH TIME	TODAY'S FLIGHT	TOTAL TIME IN SERVICE	**Description of Inspections, Tests, Repairs and Alterations** Entries must be endorsed with Name, Rating and Certificate Number of Technician or Repair Facility. (See back pages for other specific entries.)

YEAR 20____ DATE	RECORDING TACH TIME	TODAY'S FLIGHT	TOTAL TIME IN SERVICE	**Description of Inspections, Tests, Repairs and Alterations** Entries must be endorsed with Name, Rating and Certificate Number of Technician or Repair Facility. (See back pages for other specific entries.)

YEAR 20____ DATE	RECORDING TACH TIME	TODAY'S FLIGHT	TOTAL TIME IN SERVICE	**Description of Inspections, Tests, Repairs and Alterations** Entries must be endorsed with Name, Rating and Certificate Number of Technician or Repair Facility. (See back pages for other specific entries.)

YEAR 20____ DATE	RECORDING TACH TIME	TODAY'S FLIGHT	TOTAL TIME IN SERVICE	**Description of Inspections, Tests, Repairs and Alterations** Entries must be endorsed with Name, Rating and Certificate Number of Technician or Repair Facility. (See back pages for other specific entries.)

YEAR 20_____ DATE	RECORDING TACH TIME	TODAY'S FLIGHT	TOTAL TIME IN SERVICE	**Description of Inspections, Tests, Repairs and Alterations** Entries must be endorsed with Name, Rating and Certificate Number of Technician or Repair Facility. (See back pages for other specific entries.)

YEAR 20____ DATE	RECORDING TACH TIME	TODAY'S FLIGHT	TOTAL TIME IN SERVICE	**Description of Inspections, Tests, Repairs and Alterations** Entries must be endorsed with Name, Rating and Certificate Number of Technician or Repair Facility. (See back pages for other specific entries.)

YEAR 20____ DATE	RECORDING TACH TIME	TODAY'S FLIGHT	TOTAL TIME IN SERVICE	**Description of Inspections, Tests, Repairs and Alterations** Entries must be endorsed with Name, Rating and Certificate Number of Technician or Repair Facility. (See back pages for other specific entries.)

YEAR **20_____** **DATE**	**RECORDING TACH TIME**	**TODAY'S FLIGHT**	**TOTAL TIME IN SERVICE**	**Description of Inspections, Tests, Repairs and Alterations** Entries must be endorsed with Name, Rating and Certificate Number of Technician or Repair Facility. (See back pages for other specific entries.)

YEAR 20_____ DATE	RECORDING TACH TIME	TODAY'S FLIGHT	TOTAL TIME IN SERVICE	**Description of Inspections, Tests, Repairs and Alterations** Entries must be endorsed with Name, Rating and Certificate Number of Technician or Repair Facility. (See back pages for other specific entries.)

YEAR 20_____ DATE	RECORDING TACH TIME	TODAY'S FLIGHT	TOTAL TIME IN SERVICE	**Description of Inspections, Tests, Repairs and Alterations** Entries must be endorsed with Name, Rating and Certificate Number of Technician or Repair Facility. (See back pages for other specific entries.)

YEAR 20_____ DATE	RECORDING TACH TIME	TODAY'S FLIGHT	TOTAL TIME IN SERVICE	**Description of Inspections, Tests, Repairs and Alterations** Entries must be endorsed with Name, Rating and Certificate Number of Technician or Repair Facility. (See back pages for other specific entries.)

YEAR 20____ DATE	RECORDING TACH TIME	TODAY'S FLIGHT	TOTAL TIME IN SERVICE	**Description of Inspections, Tests, Repairs and Alterations** Entries must be endorsed with Name, Rating and Certificate Number of Technician or Repair Facility. (See back pages for other specific entries.)

YEAR 20_____ DATE	RECORDING TACH TIME	TODAY'S FLIGHT	TOTAL TIME IN SERVICE	**Description of Inspections, Tests, Repairs and Alterations** Entries must be endorsed with Name, Rating and Certificate Number of Technician or Repair Facility. (See back pages for other specific entries.)

YEAR 20_____ DATE	RECORDING TACH TIME	TODAY'S FLIGHT	TOTAL TIME IN SERVICE	**Description of Inspections, Tests, Repairs and Alterations** Entries must be endorsed with Name, Rating and Certificate Number of Technician or Repair Facility. (See back pages for other specific entries.)

YEAR 20____ DATE	RECORDING TACH TIME	TODAY'S FLIGHT	TOTAL TIME IN SERVICE	**Description of Inspections, Tests, Repairs and Alterations** Entries must be endorsed with Name, Rating and Certificate Number of Technician or Repair Facility. (See back pages for other specific entries.)

YEAR 20_____ DATE	RECORDING TACH TIME	TODAY'S FLIGHT	TOTAL TIME IN SERVICE	**Description of Inspections, Tests, Repairs and Alterations** Entries must be endorsed with Name, Rating and Certificate Number of Technician or Repair Facility. (See back pages for other specific entries.)

YEAR 20_____ DATE	RECORDING TACH TIME	TODAY'S FLIGHT	TOTAL TIME IN SERVICE	**Description of Inspections, Tests, Repairs and Alterations** Entries must be endorsed with Name, Rating and Certificate Number of Technician or Repair Facility. (See back pages for other specific entries.)

YEAR 20_____ DATE	RECORDING TACH TIME	TODAY'S FLIGHT	TOTAL TIME IN SERVICE	**Description of Inspections, Tests, Repairs and Alterations** Entries must be endorsed with Name, Rating and Certificate Number of Technician or Repair Facility. (See back pages for other specific entries.)

YEAR 20_____ DATE	RECORDING TACH TIME	TODAY'S FLIGHT	TOTAL TIME IN SERVICE	**Description of Inspections, Tests, Repairs and Alterations** Entries must be endorsed with Name, Rating and Certificate Number of Technician or Repair Facility. (See back pages for other specific entries.)

YEAR 20_____ DATE	RECORDING TACH TIME	TODAY'S FLIGHT	TOTAL TIME IN SERVICE	**Description of Inspections, Tests, Repairs and Alterations** Entries must be endorsed with Name, Rating and Certificate Number of Technician or Repair Facility. (See back pages for other specific entries.)

YEAR 20____ DATE	RECORDING TACH TIME	TODAY'S FLIGHT	TOTAL TIME IN SERVICE	**Description of Inspections, Tests, Repairs and Alterations** Entries must be endorsed with Name, Rating and Certificate Number of Technician or Repair Facility. (See back pages for other specific entries.)

YEAR 20_____ DATE	RECORDING TACH TIME	TODAY'S FLIGHT	TOTAL TIME IN SERVICE	**Description of Inspections, Tests, Repairs and Alterations** Entries must be endorsed with Name, Rating and Certificate Number of Technician or Repair Facility. (See back pages for other specific entries.)

YEAR 20_____ DATE	RECORDING TACH TIME	TODAY'S FLIGHT	TOTAL TIME IN SERVICE	**Description of Inspections, Tests, Repairs and Alterations** Entries must be endorsed with Name, Rating and Certificate Number of Technician or Repair Facility. (See back pages for other specific entries.)

YEAR 20_____ DATE	RECORDING TACH TIME	TODAY'S FLIGHT	TOTAL TIME IN SERVICE	**Description of Inspections, Tests, Repairs and Alterations** Entries must be endorsed with Name, Rating and Certificate Number of Technician or Repair Facility. (See back pages for other specific entries.)

YEAR 20_____ DATE	RECORDING TACH TIME	TODAY'S FLIGHT	TOTAL TIME IN SERVICE	**Description of Inspections, Tests, Repairs and Alterations** Entries must be endorsed with Name, Rating and Certificate Number of Technician or Repair Facility. (See back pages for other specific entries.)

YEAR 20____ DATE	RECORDING TACH TIME	TODAY'S FLIGHT	TOTAL TIME IN SERVICE	**Description of Inspections, Tests, Repairs and Alterations** Entries must be endorsed with Name, Rating and Certificate Number of Technician or Repair Facility. (See back pages for other specific entries.)

YEAR 20____ DATE	RECORDING TACH TIME	TODAY'S FLIGHT	TOTAL TIME IN SERVICE	**Description of Inspections, Tests, Repairs and Alterations** Entries must be endorsed with Name, Rating and Certificate Number of Technician or Repair Facility. (See back pages for other specific entries.)

YEAR 20____ DATE	RECORDING TACH TIME	TODAY'S FLIGHT	TOTAL TIME IN SERVICE	**Description of Inspections, Tests, Repairs and Alterations** Entries must be endorsed with Name, Rating and Certificate Number of Technician or Repair Facility. (See back pages for other specific entries.)

YEAR 20____ DATE	RECORDING TACH TIME	TODAY'S FLIGHT	TOTAL TIME IN SERVICE	**Description of Inspections, Tests, Repairs and Alterations** Entries must be endorsed with Name, Rating and Certificate Number of Technician or Repair Facility. (See back pages for other specific entries.)

YEAR 20_____ DATE	RECORDING TACH TIME	TODAY'S FLIGHT	TOTAL TIME IN SERVICE	**Description of Inspections, Tests, Repairs and Alterations** Entries must be endorsed with Name, Rating and Certificate Number of Technician or Repair Facility. (See back pages for other specific entries.)

YEAR 20____ DATE	RECORDING TACH TIME	TODAY'S FLIGHT	TOTAL TIME IN SERVICE	**Description of Inspections, Tests, Repairs and Alterations** Entries must be endorsed with Name, Rating and Certificate Number of Technician or Repair Facility. (See back pages for other specific entries.)

YEAR 20____ DATE	RECORDING TACH TIME	TODAY'S FLIGHT	TOTAL TIME IN SERVICE	**Description of Inspections, Tests, Repairs and Alterations** Entries must be endorsed with Name, Rating and Certificate Number of Technician or Repair Facility. (See back pages for other specific entries.)

YEAR 20_____ DATE	RECORDING TACH TIME	TODAY'S FLIGHT	TOTAL TIME IN SERVICE	**Description of Inspections, Tests, Repairs and Alterations** Entries must be endorsed with Name, Rating and Certificate Number of Technician or Repair Facility. (See back pages for other specific entries.)

YEAR 20____ DATE	RECORDING TACH TIME	TODAY'S FLIGHT	TOTAL TIME IN SERVICE	**Description of Inspections, Tests, Repairs and Alterations** Entries must be endorsed with Name, Rating and Certificate Number of Technician or Repair Facility. (See back pages for other specific entries.)

YEAR 20____ DATE	RECORDING TACH TIME	TODAY'S FLIGHT	TOTAL TIME IN SERVICE	**Description of Inspections, Tests, Repairs and Alterations** Entries must be endorsed with Name, Rating and Certificate Number of Technician or Repair Facility. (See back pages for other specific entries.)

YEAR 20____ DATE	RECORDING TACH TIME	TODAY'S FLIGHT	TOTAL TIME IN SERVICE	**Description of Inspections, Tests, Repairs and Alterations** Entries must be endorsed with Name, Rating and Certificate Number of Technician or Repair Facility. (See back pages for other specific entries.)

YEAR 20____ DATE	RECORDING TACH TIME	TODAY'S FLIGHT	TOTAL TIME IN SERVICE	**Description of Inspections, Tests, Repairs and Alterations** Entries must be endorsed with Name, Rating and Certificate Number of Technician or Repair Facility. (See back pages for other specific entries.)

YEAR 20_____ DATE	RECORDING TACH TIME	TODAY'S FLIGHT	TOTAL TIME IN SERVICE	**Description of Inspections, Tests, Repairs and Alterations** Entries must be endorsed with Name, Rating and Certificate Number of Technician or Repair Facility. (See back pages for other specific entries.)

YEAR 20____ DATE	RECORDING TACH TIME	TODAY'S FLIGHT	TOTAL TIME IN SERVICE	**Description of Inspections, Tests, Repairs and Alterations** Entries must be endorsed with Name, Rating and Certificate Number of Technician or Repair Facility. (See back pages for other specific entries.)

YEAR 20____ DATE	RECORDING TACH TIME	TODAY'S FLIGHT	TOTAL TIME IN SERVICE	**Description of Inspections, Tests, Repairs and Alterations** Entries must be endorsed with Name, Rating and Certificate Number of Technician or Repair Facility. (See back pages for other specific entries.)

YEAR 20_____ DATE	RECORDING TACH TIME	TODAY'S FLIGHT	TOTAL TIME IN SERVICE	**Description of Inspections, Tests, Repairs and Alterations** Entries must be endorsed with Name, Rating and Certificate Number of Technician or Repair Facility. (See back pages for other specific entries.)

YEAR 20_____ DATE	TOTAL TIME IN SERVICE	**Reference of Major Repairs and Major Alterations To** FAA Form 337 by Date, or to the Work Order by Number and the Approving Agency

YEAR 20_____ DATE	TOTAL TIME IN SERVICE	**Reference of Major Repairs and Major Alterations To** FAA Form 337 by Date, or to the Work Order by Number and the Approving Agency

YEAR 20____ DATE	TOTAL TIME IN SERVICE	**Reference of Major Repairs and Major Alterations To** FAA Form 337 by Date, or to the Work Order by Number and the Approving Agency

YEAR 20_____ DATE	AD NUMBER	TOTAL TIME IN SERVICE	**Airworthiness Directives** Chronological Listing and Method of Compliance

YEAR 20____ DATE	AD NUMBER	TOTAL TIME IN SERVICE	**Airworthiness Directives** Chronological Listing and Method of Compliance

YEAR 20_____ DATE	AD NUMBER	TOTAL TIME IN SERVICE	**Airworthiness Directives** Chronological Listing and Method of Compliance

YEAR 20_____ DATE	TOTAL TIME IN SERVICE	**Manufacturers' Mandatory Service Bulletins** Chronological Listing and Method of Compliance

YEAR 20_____ DATE	TOTAL TIME IN SERVICE	**Equipment Addition, Removal or Exchange** Item Manufacturer's Name Model Serial Number	
			☐ Addition of Optional Equipment ☐ Removal of Optional Equipment ☐ Addition of Required—Exchanged for Optional ☐ Removal of Required—Exchanged for Optional
			☐ Addition of Optional Equipment ☐ Removal of Optional Equipment ☐ Addition of Required—Exchanged for Optional ☐ Removal of Required—Exchanged for Optional
			☐ Addition of Optional Equipment ☐ Removal of Optional Equipment ☐ Addition of Required—Exchanged for Optional ☐ Removal of Required—Exchanged for Optional
			☐ Addition of Optional Equipment ☐ Removal of Optional Equipment ☐ Addition of Required—Exchanged for Optional ☐ Removal of Required—Exchanged for Optional
			☐ Addition of Optional Equipment ☐ Removal of Optional Equipment ☐ Addition of Required—Exchanged for Optional ☐ Removal of Required—Exchanged for Optional
			☐ Addition of Optional Equipment ☐ Removal of Optional Equipment ☐ Addition of Required—Exchanged for Optional ☐ Removal of Required—Exchanged for Optional

YEAR 20____ DATE	TOTAL TIME IN SERVICE	**Equipment Addition, Removal or Exchange** Item Manufacturer's Name Model Serial Number	
			☐ Addition of Optional Equipment ☐ Removal of Optional Equipment ☐ Addition of Required—Exchanged for Optional ☐ Removal of Required—Exchanged for Optional
			☐ Addition of Optional Equipment ☐ Removal of Optional Equipment ☐ Addition of Required—Exchanged for Optional ☐ Removal of Required—Exchanged for Optional
			☐ Addition of Optional Equipment ☐ Removal of Optional Equipment ☐ Addition of Required—Exchanged for Optional ☐ Removal of Required—Exchanged for Optional
			☐ Addition of Optional Equipment ☐ Removal of Optional Equipment ☐ Addition of Required—Exchanged for Optional ☐ Removal of Required—Exchanged for Optional
			☐ Addition of Optional Equipment ☐ Removal of Optional Equipment ☐ Addition of Required—Exchanged for Optional ☐ Removal of Required—Exchanged for Optional
			☐ Addition of Optional Equipment ☐ Removal of Optional Equipment ☐ Addition of Required—Exchanged for Optional ☐ Removal of Required—Exchanged for Optional

Notes

Notes